AMAZING ADVENTURES ON LAND

WRITTEN BY
ALEX HALL

All rights reserved.
Printed in India.

A catalogue record for this book is available from the British Library.

ISBN: 978-1-80505-600-3

Written by:
Alex Hall

Edited by:
Rebecca Phillips-Bartlett

Designed by:
Jasmine Pointer

©2024
BookLife Publishing Ltd.
King's Lynn, Norfolk
PE30 4LS, UK

All facts, statistics, web addresses and URLs in this book were verified as valid and accurate at time of writing. No responsibility for any changes to external websites or references can be accepted by either the author or publisher.

AN INTRODUCTION TO BOOKLIFE RAPID READERS...

Packed full of gripping topics and twisted tales, BookLife Rapid Readers are perfect for older children looking to propel their reading up to top speed. With three levels based on our planet's fastest animals, children will be able to find the perfect point from which to accelerate their reading journey. From the spooky to the silly, these roaring reads will turn every child at every reading level into a prolific page-turner!

CHEETAH
The fastest animals on land, cheetahs will be taking their first strides as they race to top speed.

MARLIN
The fastest animals under water, marlins will be blasting through their journey.

FALCON
The fastest animals in the air, falcons will be flying at top speed as they tear through the skies.

CONTENTS

Page 4	Your Journey on Land
Page 6	Marco Polo
Page 8	Sacagawea
Page 12	David Livingstone
Page 14	Nellie Bly
Page 18	Gertrude Bell
Page 20	Annie Londonderry
Page 24	Roald Amundsen
Page 26	Edmund Hillary and Tenzing Norgay
Page 30	Where Will Your Journey on Land Take You?
Page 31	Glossary
Page 32	Index

Words that look like THIS are explained in the glossary on page 31.

Your Journey on Land

Welcome, adventurers! We are going to follow the paths of some of the most amazing land explorers.

Many great explorers have travelled through forests, across deserts and up mountain tops to learn more about the world.

There is still so much left to explore. Are you ready to begin our journey on land?

Then what are you waiting for? Get your hiking boots on and let the adventure begin!

MARCO POLO

1254–1324

Our journey starts with Marco Polo in Italy. Polo left Italy with his father and uncle to go on a long journey to China.

The <u>emperor</u> of China wanted to learn more about Europe. Polo wanted to learn about China.

6

The emperor gave Polo a golden passport to travel wherever he wanted. He travelled a lot of the trip by camel.

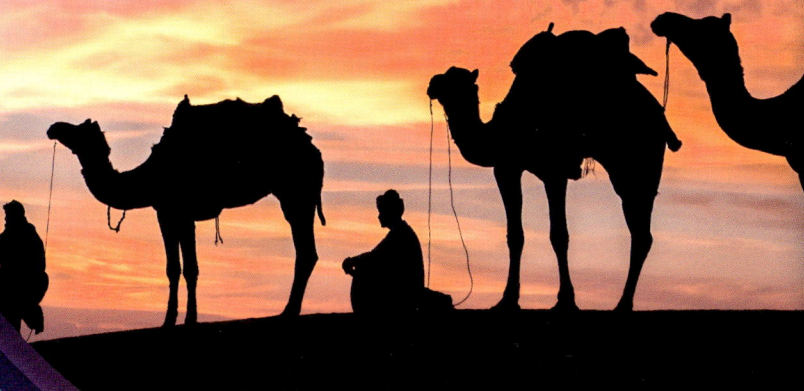

Polo wrote a book about his travels. This helped Europeans understand what life was like in Asia.

SACAGAWEA
Around 1788–around 1812

It is time to travel across the United States. We are following Sacagawea, a Native American woman from the Shoshone tribe.

Sacagawea became the guide to some explorers called Meriwether Lewis and William Clark. They were on an expedition across America.

When Sacagawea joined the expedition, she had just had a baby. She carried and cared for her newborn son throughout the entire journey.

Sacagawea also helped to find food for everyone. She knew which plants were safe to eat.

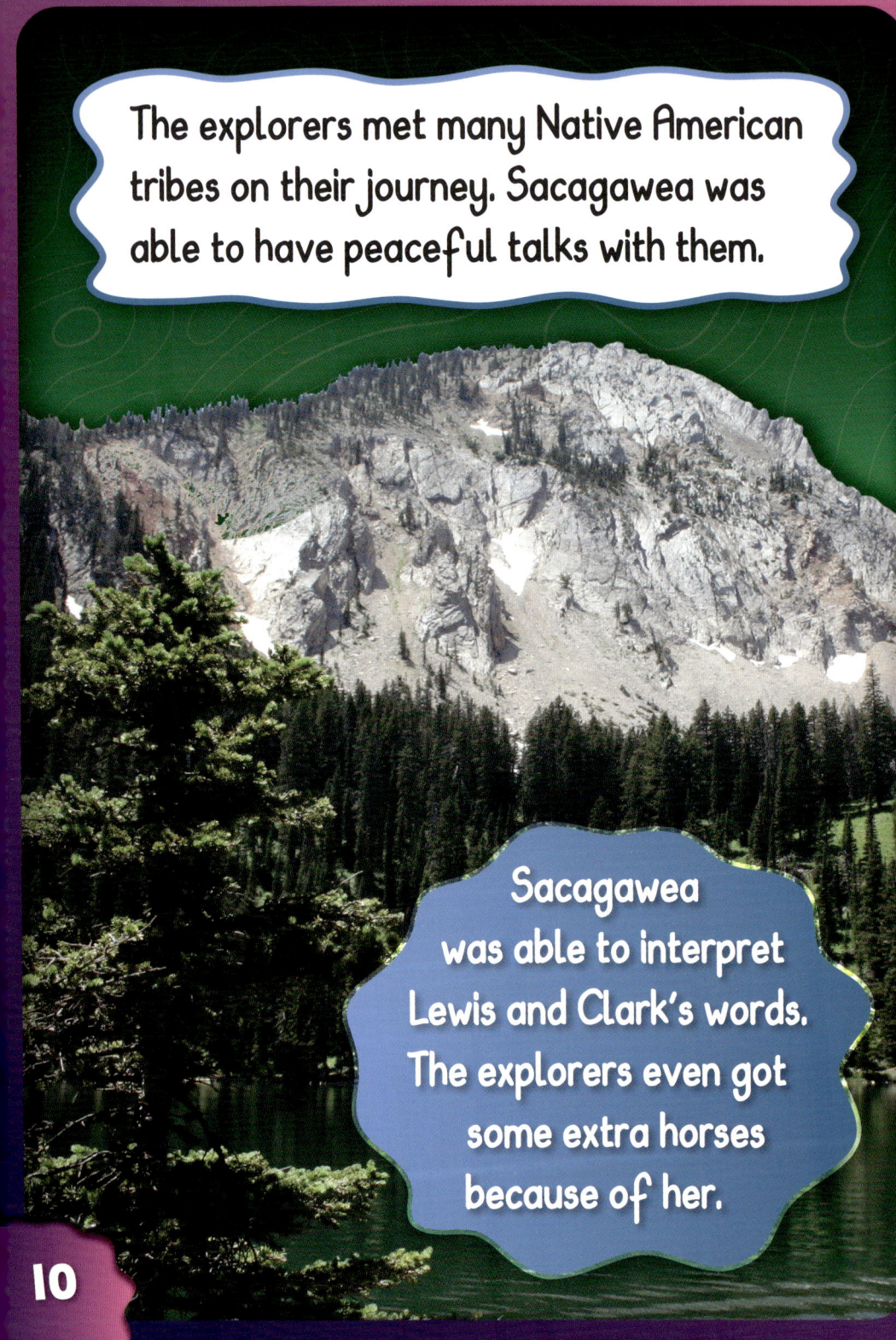

The explorers met many Native American tribes on their journey. Sacagawea was able to have peaceful talks with them.

Sacagawea was able to interpret Lewis and Clark's words. The explorers even got some extra horses because of her.

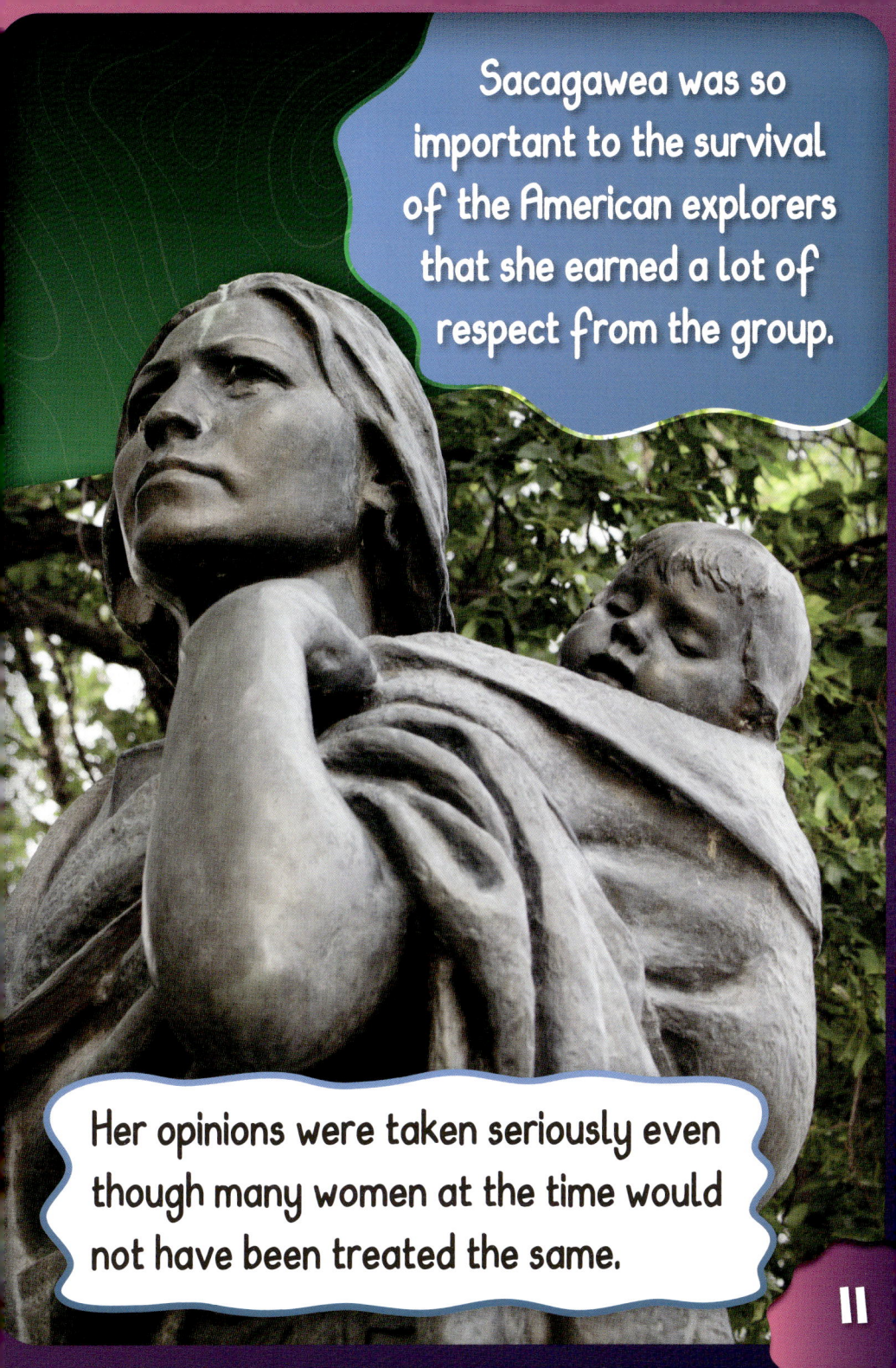

Sacagawea was so important to the survival of the American explorers that she earned a lot of respect from the group.

Her opinions were taken seriously even though many women at the time would not have been treated the same.

DAVID LIVINGSTONE
1813–1873

We are now following Scottish doctor David Livingstone. He travelled across Africa and fought the slave trade there.

He thought that if he could find new trade opportunities by exploring Africa, it could replace the practice of selling enslaved people.

Livingstone disappeared while searching for the start of the Nile, the world's longest river. He was eventually found alive.

He did not find the start of the Nile. However, many of his discoveries helped replace the slave trade in East Africa.

Nellie Bly

1864-1922

The next part of our journey begins in New York, where Nellie Bly was born.

Nellie Bly was a <u>journalist</u>. Her job was to write important stories to teach people more about the world.

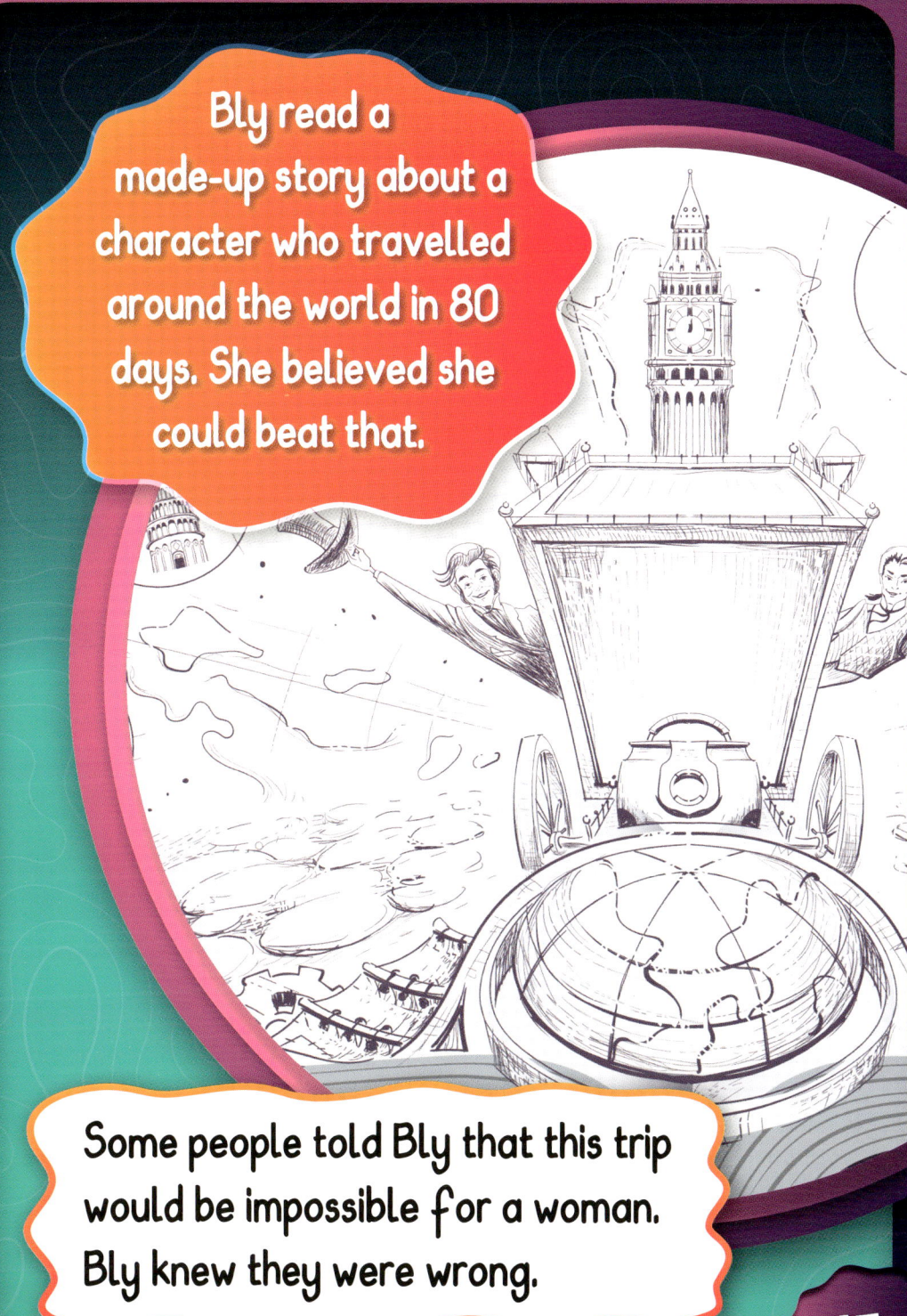

Bly read a made-up story about a character who travelled around the world in 80 days. She believed she could beat that.

Some people told Bly that this trip would be impossible for a woman. Bly knew they were wrong.

Bly set out on her journey. She used trains, ships and even donkeys to get around the world.

Bly made it all the way around the world in just 72 days. Everyone was shocked to see her return so quickly.

News of Bly's trip made her famous. She wrote a book about her journey around the world.

Nellie Bly always challenged herself to step into the unknown. That determination helped her break records and be remembered in history.

GERTRUDE BELL
1868-1926

If you want to see a fearless adventurer, then look no further. Gertrude Bell was a British climber and <u>archaeologist</u>.

Bell climbed many mountains in the Alps. She even has a mountain peak named after her.

During one climb, Bell got <u>frostbite</u>. This did not stop her love for climbing. Just two years later she climbed a mountain called Matterhorn.

Bell was often the only woman around. It never stopped her from achieving amazing things.

MATTERHORN

Annie Londonderry

1870-1947

Annie Londonderry moved to the United States at a young age. When she was about 23, she decided to go around the world on a bicycle.

No woman had ever done it before, but she wanted to try.

When Londonderry started, she struggled because of her heavy bicycle and long skirt. They made the journey slow and difficult.

BLOOMERS

Londonderry did not give up. She switched to a lighter bicycle and trousers called bloomers to make things easier.

Londonderry finished her journey in a suit that was made for a man.

Londonderry completed her journey in 15 months. She did use some boats and trains on the way, but she had done something that many thought was impossible.

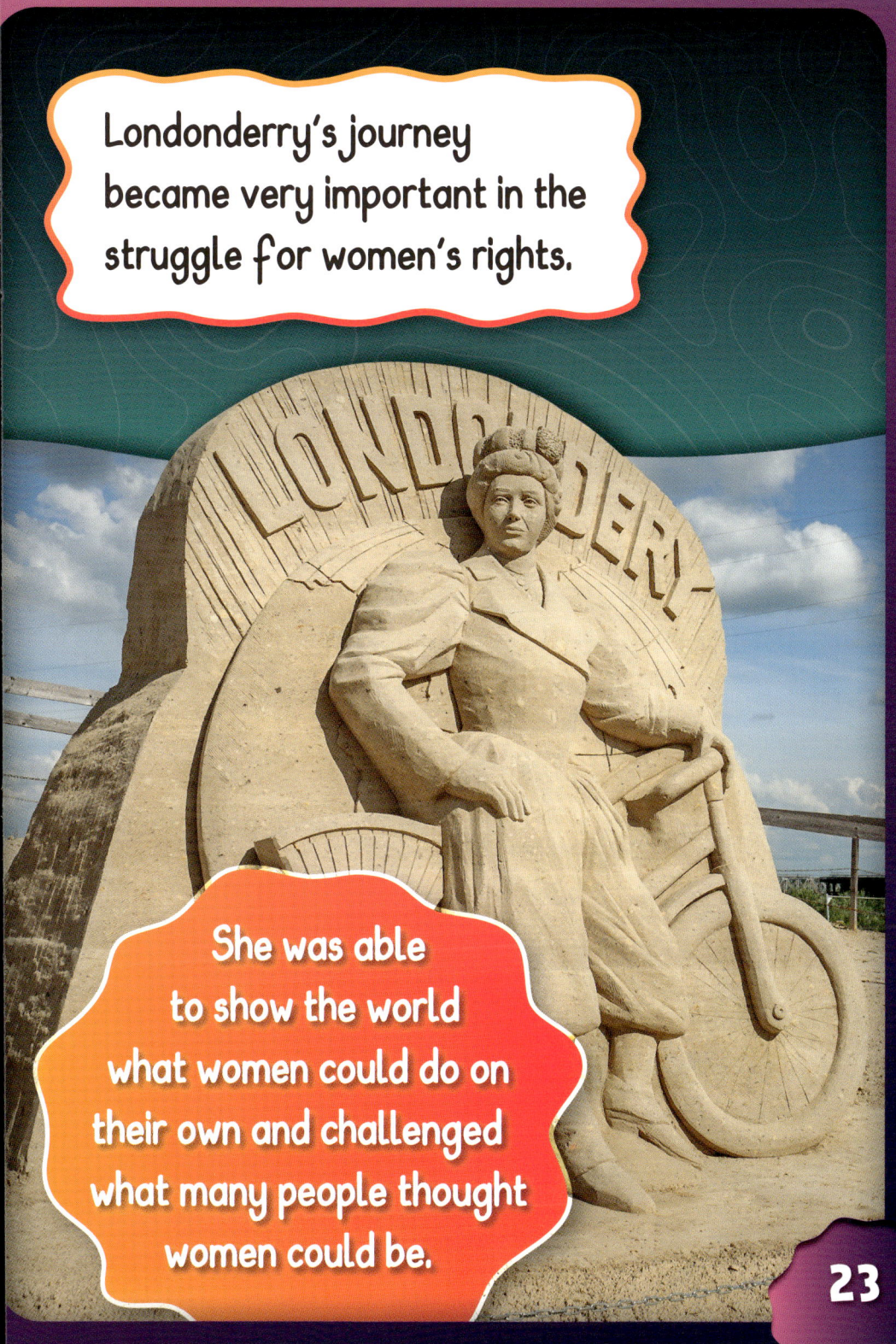

Londonderry's journey became very important in the struggle for women's rights.

She was able to show the world what women could do on their own and challenged what many people thought women could be.

ROALD AMUNDSEN
1872-1928

Now we are going to the South Pole, the most southern place on Earth. It was the goal of Norwegian man, Roald Amundsen, to reach it.

Others tried before him, but the cold weather made it a difficult journey.

The weather meant that Amundsen needed to prepare. He dressed in warm clothes and used <u>dog sleds</u> to go over ice.

He became the first person to reach the South Pole. With careful preparation, anything is possible.

Edmund Hillary
and 1919-2008
Tenzing Norgay
1914-1986

Let us finish on top. We are following Edmund Hillary and Tenzing Norgay in their goal to climb the highest mountain in the world, Mount Everest.

The New Zealander and the Sherpa were chosen to make the climb together.

Many other people tried to climb Everest. However, bad weather, the freezing cold temperatures and the lack of air made them unsuccessful.

MOUNT EVEREST

Just three days after a previous team failed, Hillary and Tenzing Norgay started to climb.

This trip was Hillary's fourth expedition in the Himalayas in just over two years. It was Tenzing Norgay's seventh attempt to climb Mount Everest.

Hillary and Tenzing Norgay's combined climbing experience made them a great team.

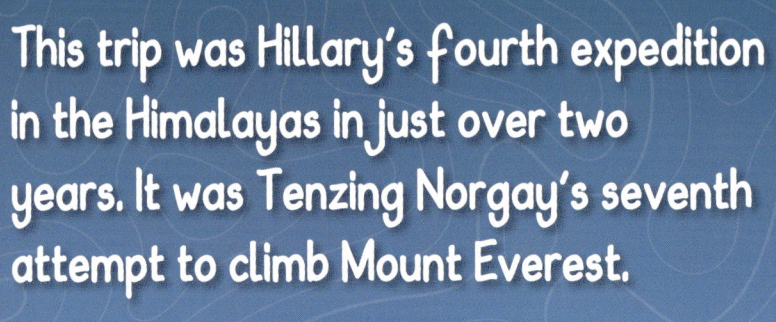

They finally reached the top. However, they did not stay long because they only had 15 minutes of air left.

Hillary and Tenzing Norgay were the first people to climb Mount Everest. Their teamwork and skills helped them climb to fame.

Where will your journey on land take you?

We have reached the end. You should be proud that you got here. You must need a rest after all those adventures.

There are still discoveries on land to be made. Go out and see what you can find!

GLOSSARY

archaeologist   someone who finds and studies things hidden in the ground, such as fossils and other things of importance from the past

dog sleds   small, flat vehicles that are pulled along by a group of dogs

emperor   a person who rules a country or group of countries

expedition   a journey for a specific purpose

frostbite   when a living thing gets affected or damaged by extreme cold

journalist   someone who researches and writes news stories for people to read

Sherpa   a group of people who live near the area around the Himalayan mountains, who are known for their mountain climbing abilities

slave trade   the business of capturing and selling human beings

tribe   a group of people who have the same language and ways of doing things

INDEX

Africa 12–13

Alps 18

America 8

books 7, 17

mountains 4, 18–19, 26

rivers 13

Shoshone 8

South Pole 24–25

trains 16, 22

world, the 4, 13–17, 20, 23, 26

Photo Credits

Images are courtesy of Shutterstock.com. With thanks to Getty Images, Thinkstock Photo and iStockphoto.
Recurring images — Dancake. Cover — Rudra Narayan Mitra, Vixit, Dancake. 4–5 — Charles Knowles, Winston Springwater. 6–7 — Salviati, Public domain, via Wikimedia Commons, saisnaps. 8–9 — Edgar Samuel Paxson, Public domain, via Wikimedia Commons, Dsdugan, CC0, via Wikimedia Commons. 10–11 — Tracy Grazley, Ace Diamond. 12–13 — Everett Collection. 14–15 — H. J. Myers, photographer, Public domain, via Wikimedia Commons, ALEXEY GRIGOREV. 16–17 — Wm. Notman and Son, Public domain, via Wikimedia Commons, Johnston, John S., Public domain, via Wikimedia Commons, Melnikov Dmitriy, Unknown author, collection of the Museum of the City of New York via Wikimedia Commons. 18–19 — picture copied from the Gertrude Bell Archive (1), Public domain, via Wikimedia Commons, Jakl Lubos. 20–21 — Studio Tourne, Public domain, via Wikimedia Commons, KathySG. 22–23 — William J. Root, Chicago, Public domain, via Wikimedia Commons, Regina M art. 24–25 — AnonymousUnknown author, Public domain, via Wikimedia Commons, bilinmiyor, CC BY-SA 4.0 <https://creativecommons.org/licenses/by-sa/4.0>, via Wikimedia Commons. 26–27 — Jamling Tenzing Norgay, CC BY-SA 3.0 <https://creativecommons.org/licenses/by-sa/3.0>, via Wikimedia Commons, Vixit. 28–29 — Sergey Goryachev. 30 — solarseven.